The prisoners of Comfort

Vitamins

by
Jim Plagakis, RPh

"In a hole in the ground there lived a hobbit. Not a nasty, dirty, wet hole, filled with the ends of worms and an oozy smell, nor yet a dry, bare, sandy hole with nothing in it to sit down on or to eat: It was a hobbit-hole, and that means comfort."

John Ronald Reuel Tolkien

The Prisoners of Comfort

Copyright by Jim Plagakis 2010

      For twenty years Jim Plagakis has been the author of the popular Drug Topics Magazine column JP at Large. He has been a prescient observer of the drug store industry and his predictions have been consistently accurate. Jim loves pharmacy and he knows that it is the job that can present problems. The profession is just fine.

      Mister Plagakis was one of the first to identify institutionalization of the pharmacist in the modern chain drug store culture. This social phenomenon is rather recent and became pronounced beginning in the early 1990s when the industry began depending on computer programs for productivity. Reports became the rulers of the pharmacist's behavior and prescription counts became the dictator of their time.   Moore's Law regards exponential increases. Pharmacists filled around $30 Billion worth of Rx in 1989. $105 billion worth of Rx in 1999. $234 billion in 2008. How long will it take to get to a trillion dollars? Moore's Law says not very long. It is no wonder that keeping the Prescription Mills grinding is so important to the big drug store chains.

The price is that highly educated medical professionals are asked to dumb down and toe the mark.

Institutionalization

To put someone into an institution such as a prison.

To make something an established custom or accepted part of the (job)

A good example is simply going to the bathroom when you have to go. Pharmacists will consistently answer one more phone call, check one more prescription and answer one more question. Often, they make it to the bathroom at the crisis stage and end up with damp underwear.

The Players

Pharmacists are The Prisoners

The Non-Pharmacist Managers are the Jailers

The Middle-Managers (District) are the Wardens

The Executives are the Governors

Pharmacy Culture is: The Prison

I am an institutionalized pharmacist. I have been this way so long that it is part of who I am. I am used to eating my meals at work standing up. I am used to being interrupted by the telephone, a patient with a question or any of a number of other reasons that my meal gets cold if I am stupid enough that day to bring something that is better warm rather than my usual cold sandwich. I drink too many caffeinated beverages and I succumb to Snickers bars, salty chips and cookies. In the past 43 years, I may have actually taken a real lunch, when I could leave the store, a couple hundred times. That's only six times a year. It did not take me very long in 1965 to get with the program and surrender to the established lunch time custom in pharmacy.

My legs are shot. I have worked so many twelve and thirteen hour days without a break that it is amazing that I can still stand for eight hours. I am so used to standing all day long that I even stand when it is not busy and the stool is right beside me. I actually feel uncomfortable sitting down at work. I have Post-Polio Syndrome and concomitant neuropathy in my legs. I can't blame the pharmacy culture for that, but I have not used good judgment. Pharmacists stand and they stand all day long. That's the culture so that is what I did and I am still doing it forty three years after the starting gun went off.

Pharmacies are notoriously poorly designed for the health of pharmacists. The computer screens are fixed at a certain height. What is appropriate for a man six foot two is not right for a girl five foot four. The floors at too many places have no comfort pads or cheap imitations. Holding the telephone between your ear and shoulder can cause painful conditions in your neck. In January, 2010, I turned myself over to two Physical Therapists. I had terrible neck and shoulder pain from a head-forward posture after four decades of bending over the counter. It took almost a year, but I have followed the exercise program religiously and stand straight again, like when I was a boy. Why did I put up with it for 40 years?

I believe that pharmacists should be managed by pharmacists. I do not believe that it is appropriate for a highly educated medical professional to be managed by a non-pharmacist manager who may or may not have any education beyond high school. I know that that sounds elitist, but so be it. Non-Pharmacist Managers are the Jailers in our pharmacy culture. They are more often than not products of the culture. For the Non-Pharmacist Manager, company rules trump the pharmacist's personal ethics, professional standards and legal responsibilities. The Jailers enforce the company's rules and the mandates for completing reports on time. The Jailers are particularly zealous about the Prescription Mill. Some companies have timers that report how fast the pharmacist runs The Prescription Mill via computer-generated reports. If you lag behind in the culture of some Chain Drug Stores, you are brought onto the carpet, even written up.

There have been some Non-Pharmacist Managers who could not stand me. They did not like that I did not do what they wanted me to do just because they were ordering me to do it. If it was a professional matter, the Non-Pharmacist Manager had absolutely no standing. If it was a pharmacy business matter, I would consistently make my own choices or confer with The Warden, the Pharmacist District Manager. A few Jailers absolutely hated me. They hated that I did not work as many hours as they did and that I earned more money than they did. They were unhappy managers and that made them even more aggressive about making me abide by the rules of The Prison. Sometimes I did and sometimes I did not. I was one of those pharmacist who the Jailers wanted in "the hole". A disclaimer. I have worked with Non-Pharmacist Managers who were happy Jailers. They recognized that I was a talented Pharmacy Manager and that what I did made their bonus checks fatter.

Store managers can be petty, mean and spiteful. They look for every chance to criticize your behavior. They love to write the Prisoners up. I had a non-pharmacist manager who insisted that I sign up on the calendar for bathroom cleaning detail. I was a janitor for awhile in college so I know how to clean. When my day came, I took the vacuum, mop and pail, electric buffer, Windex, sink cleaner and other assorted supplies into the men's room. I was having a merry time cleaning. It was a terrific break from The Prescription Mill. After only twenty minutes, The store manager burst into the men's room. He demanded to know what I was doing. I reminded him of the cleaning schedule. He was so angry there was a spray of spit as he shouted at me, demanding that I get back to the pharmacy. Another manager wrote me up because I refused to wear the black slacks and white shirts that were company policy. I reminded him that Washington State law required the company to purchase any clothes that were required for the job. I had a manager tell me that I couldn't chew my food in sight of the patients. Another criticized me because I have a habit of complimenting women when they look good. He said it could be construed as sexual harassment. The Jailers will always look for weaknesses to exploit. I learned to either never show weakness or to stand my ground. I am a Prisoner, but I am a medical professional Prisoner.

Vacations are one of the benefits that the companies ballyhoo when you are first looking at the job. Most drug store chains promise multiple weeks of vacation after only one year of service. I don't think that I have enjoyed vacations in the summer more than a half dozen times. Even after years of service, they made it very difficult to plan a vacation when the kids were out of school. The Jailers often had a little grin when they told me, "Plagakis, you didn't get those two weeks in July that you wanted. I told them to put you down for one week in August and the other in November."

One week in 1976, I worked sick. There was a virus in the community and every employee of the little drug store I managed was out sick with diarrhea and nausea and vomiting. I was the manager. My Saturday relief was not available during the week so I worked sick. I had to make a sign that read "Be back in a minute". I had to use the bathroom often for a couple days. I would lock the front door and put up the sign. I have worked sick too many times because that is what the culture expects from a pharmacist. The managers figured this out quickly. They always expected me to work while I was ill.

The word "Rude" is a buzzword for The Jailers and The Wardens. Smart rat drug store customers know that if they use that word the Jailer will be all over the Prisoner. I have been accused of being rude countless times and almost every time, I have been put in the corner with the dunce cap on my head. I have been written up because customers have accused me of being rude. The Jailers seemed to relish the moments in the office when I was obliged to give my side of the story. Occasionally, I was written up. The Jailer was always furious when I refused to sign the write up form. Prisoners aren't allowed to do that. I still did. My manner may be abrupt and business-like, but I am never rude.

The Jailer will chortle with glee when he sees The Prisoner working in the pharmacy after it has closed. The order needs to be placed. The order needs to be put up. At closing, there were still thirty prescriptions to complete. Working off-the-clock is illegal in most states if you are an hourly employee. If you are a salaried employee and your pay is based on a forty hour week and you actually work sixty hours, most state laws require that you be paid. If a pharmacist is in the habit of working off-the-clock, they must document every single hour, including the reason why they stayed late or came in early. If the state doesn't have progressive laws, the federal government does.

The Prisoner will often have to work without enough help. There will be times when The Prisoner is in a situation where he will have to run The Prescription Mill, answer the telephone, attend to the Drive Through and be friendly and forthcoming at the main counter. The law as well as personal standards and professional ethics require that he counsel patients on new prescriptions. I have been in that situation numerous times. I have closed the Drive Through with the excuse that there is an electrical problem. When The manager found out, he lambasts me. I listen carefully like a good Prisoner, but I do not promise to never do it again. The telephone is a last priority. I just do not answer it. I bounce back and forth between The Prescription Mill and the main counter. I do my best and my number one rule is everyone has to wait. The Governors are at fault when there is not enough help. The Governors care only about numbers. Primarily the bottom line. They are the ones who sign contracts with the PBMs that put the company at a disadvantage right out of the starting gate. The profits are pitifully thin. The Governors then go to the Wardens and demand to see better results. The Wardens then go where they always go. They go to The Jailers and demand that the pharmacy payroll be pared down. It is always technicians who lose hours and The Prisoners have to work without enough help. Every pharmacist knows that it is customer service that is affected negatively. The Jailer doesn't care. It is just one more chance to blame The pharmacist for something that is not their fault.

Just about every pharmacist has professional standards that are not being fulfilled. It bothers them that they keep their noses in The Prescription Mill and that they do not live up to their obligations to provide superior patient care by counseling when appropriate. It bothers them that they do not have the time to give personal attention to patients who need assistance with OTC items. They know that OTC products are drugs and just as dangerous as Rx Only products. Some chain drug stores have timers. At least one chain is driven by the timers. The Prisoner is on a "chain gang". Unfortunately for the profession, there are pharmacists who have given up. They view their situation as Lifers with no possibility of parole. They accept the shackles. They trudge along. They are happy to collect their handsome paychecks, but make no attempt to make the choices that could change their lives. Worst of all, they blame the profession.

Being a Prisoner is not a position in which a person can thrive. There is little that a pharmacist can do to become a self-actualized person. The Prisoner's life is a set of rules. Some Prisoners rebel and overreact aggressively and quit or get fired. I quit a management job in anger once. I had worked for the company for over fifteen years. My commute was ten minutes when the traffic was bad. I felt that I was right in my dispute with a store manager. I was angry and indignant and this is a very dangerous stew. They may use food as a way out and become food addicts. They can engage in drinking too much or risky sexual habits that are often with a partner who is not a spouse. A female pharmacist was treating a vaginal infection with Metronidazole. She was an attractive petite blonde. I will always remember her perfume. She wore it tastefully, Pleasures by Elizabeth Arden. I called her a friend in arms, a sister pharmacist. She was a church-going woman with three children. She could not imagine how she was infected. I suggested that her husband could have been the culprit and she confessed that she and her husband no longer had sex. A few weeks later, the infection was back. Again, she was clueless until I asked her if she had a boyfriend. It was too much for her. She got weepy and asked me why I didn't mind my own business. I lost a friend. Soon after that, she changed jobs, then she changed towns, then she changed families. I didn't keep track. I don't even know her Christmas card address. The last I knew for sure was that she was living alone, without her three children.

Drug abuse is much more prevalent than you would think. A 2001 Drug Topics Magazine published a study by Dean Dabney, Ph.D. Georgia State University. Criminal Justice Department. 45% of Pharmacists have diverted potentially addictive pharmaceutical drugs for their own use. Near seven o'clock when the pharmacist is all alone and there are cars at the drive-through, customers standing in line at the register, six prescriptions to be filled and the phone ringing, it can be tempting to use a readily available chemical to help in getting through the day.

Other Prisoners get depressed and are barely able to function. There is a delightful artist in Seattle who probably could not survive if she identified herself as a pharmacist. For her, the Prison is so tyrannical and unfair, that her friends do not even know what she does for her living. She works a part time schedule in a chain drug store. She collects her paycheck. Pharmacy is in an armored compartment of her mind that she rarely visits.

Some pharmacists manage to remain brave, dignified and unselfish, but The Prison is not a place for courageous behavior. For many pharmacists, the conditions of the pharmacy are so oppressive that they give up. The mental conflicts are much too stressful. The clashes of will-power that they experience when they stand up to The Jailer are soul-destroying and they shrink into a true Prisoner's existence. An existence that does not even resemble the life of a highly trained medical professional.

When you are thirty years old and earn ten thousand dollars a month, you realize that your family does not have to wait for the good life. You can have it now. Making payments on your student loan at a thousand a month is not a problem. You have two nice cars, your children go to a private school, you can afford the best dance classes for your daughter and your son is already a black belt in Taekwondo. Your house is a dream that you and your spouse believed would come someday. Why wait? You told your spouse that you could have it now. The mortgage is hefty, but you can make the payments and your sisters are envious that you got that place on the lake when they may always be stuck in town. You have a life of comfort. You like it and you don't ever want to have money be a problem again. You and your spouse scrimped during the college days and you have vowed to never do it again. You may not know it, but you are a Prisoner of Comfort.

The Governors and The Wardens depend on that. They need for you to be a Prisoner. As long as Comfort is King, you may always be The Prisoner and the Pharmacy will always be your Prison.

I worked with a pharmacist in Washington State for whom there would never be enough. This man was a Prisoner of the highest order. He worked a full time job for a chain drug store and he worked weekends for a major grocery chain. He made a lot of money. He had nice things and lived in a condominium in the Cascade Mountains foothills. There was a hillside outside his living room. He could sit on his deck and hear nothing but the wind and the birds. His problem was that he didn't visit his deck very often because he was in Prison. He said that it didn't matter that his working day was boring and repetitive and that it drained his energy worse than if he had been on an actual chain gang. He declined our dinner invitation so many times that I stopped inviting him. The girls we invited were nonplussed. We said it wasn't them. He was comfortably not engaged in life.

When you are in Prison, there are forces beyond your control that can take away your freedom. That is what The Prisoner believes, however you can always choose what you feel and what you do about what happens to you. You have a choice to remain brave, dignified and unselfish, or become no more than a robot.

I worked with a pharmacist in the late 1960s. We were the same age. We were in our late twenties. It was in a union shop. We took divergent paths. I was not satisfied with the everyday sameness of the job. I wanted to exercise my management muscles so I quit the union job and began a life of looking for more. I never did find enough. My friend, on the other hand, was content. He stuck with the union job until he retired. He was a dignified man and his self-respect was not dependent on the opinion of others. A union job is positively different, however. There are protections in place that make running a pharmacy like a prison very difficult. When you complain to the union, things get done.

A pharmacist's wage allows for pleasure in life. We make enough money to do the things that bring enjoyment to our everyday existence. There is some gratification in eating good food. The little bar has only top shelf libations. Our family is safe and protected in the home our wage has provided. We wear nice clothes and have all of the amenities of the 21st Century. We may belong to social organizations and we participate in amusing recreation. For most of us, this is probably not enough. Someone did say, "You can't buy happiness".

Right out of the pot, I wanted some power. I believed that a management career would allow me to exact a revolution of sorts. It all came down to money, of course. When I produced a gross profit better than any other manager in the company, I was rewarded with a transfer from California to Whidbey Island, north of Seattle. I relished in that kind of power for about fifteen minutes. I had the terrific place to raise the kids, but I found out very quickly that, in Washington State, I was more a prisoner than ever before. I now know that the perception that one has power as a pharmacy manager is false. I actually managed the pharmacy in California. I made decisions that affected the profitability of the department. I had no latitude in Oak Harbor to do anything but let the computer make the decisions. I was suddenly and completely institutionalized. I was powerless.

I want my life to mean something. My need for pleasure is simple. There is nothing better than a cup of coffee and the company of my wife or a good friend. Of course, I like it better when I am outside in a beautiful place, preferably beside a body of water. The most exquisite pleasures I remember from years ago are always like that. Simple and uncomplicated.

Power is a vaporous thing. You can observe yourself as a manager and say, "You are one powerful, Dude, man. You make all of the decisions and everybody jumps." It's still all about money. You will find out just how powerful you are when you are consistently over the payroll budget, have an overstocked pharmacy and the wait times are too long.

The question I have asked myself is: How do I insist that the job means more than the money? I found it about ten years ago. When I am getting ready for work, I say, "You will make a contribution to another human being today, Jim." And I do. We are fortunate that all of the legal ducks are lined up in our favor. We are mandated by law to counsel. The stage is set and we are the star players. We get opportunities every single day to make a difference. If we don't take those opportunities, shame on us. The Jailers have no say on this, if they are smart. This is the best, and possibly last, good chance out of The Prison. Pharmacists complain that they do not have the time. They say that they could be penalized by the Wardens if they do not make the numbers. These people are so thoroughly institutionalized that they can't see that they are only half a pharmacist, if that. There is a right way and a wrong way to practice pharmacy. In my view, a patient-centric practice is the only way.

I can't expect my job to give me meaning. I have to find meaning in my job, all by myself.

Most of us Prisoners live a very shallow existence because we are 21st Century Americans and the place where most Americans look for meaning is their job. It is classic disappointment when, early on, we realize that we are expected to pay more attention to the Prescription Mill than to the needs of our patients. As interns, we see our preceptors growing older with no spark of life in their eyes. They just want to get through the day, make the numbers and get out of there as fast as they can. The hearts are ripped out of young pharmacists when they become conscious of what they have signed up for. The Prison, with the help of the preceptors, turns young people with dreams and goals into Prisoners, ripe with energy, who can run The Mill faster than the last guy.

Happiness in a job can't come about from pure determination and hard work. Happiness and success happen as a result of dedication to a cause greater than oneself. This does not mean that your cause has to be huge, unwieldy and requiring enormous time and effort. It just means that there has to be a reason to go to work that is bigger than making a living. You cannot will success. It just happens. It comes from a well of intention that you can replenish every single day, if you want.

How does everyday work life in the pharmacy affect the dignity, self-respect and integrity of the practicing pharmacist? This is how one pharmacist handled his well-being. He was a thirty-something family man when I knew him well. His wife was an accountant who did taxes for small corporations. His little boy was given every advantage. He had security and a bright future as a pharmacy manager. He enjoyed sporting events and played tennis on his days off. He and his wife had a date once a month. Often, they would go out of town for a honeymoon-style weekend. He was still a Prisoner and there was something lacking. It was a toxic stew for a man with high personal standards. He confessed to me years later that he just wanted to be someone's hero. For him, working as a Prisoner at a Prison made him ineligible for heroism, even for his little boy.

My friend gave it all up for his ego. First, he started to drink more. That made it worse. He suffered from sleep disturbances and confessed that there was nothing in his life that brought him pleasure. During a philosophical evening powered by Pinot Noir for me and Tequila shooters for him, he claimed that he did not see any reason to continue. That bothered me. Prozac was a relatively new drug and I urged him to talk with his doctor. Instead, he went out and bought a $30,000.00 Harley Davidson.

His wife fought him. She refused to ride with him so he went and found a girl who would. He grew a biker's beard and let his hair grow. The bike and his new friends became his life. He made a joke once about swiping Vicodin, but never said a never another word about it. When I asked him, he just laughed it off. I contacted him a few months ago. It was twenty years later. This fifty-something man is single. His twenty-something son avoids him. His ex-wife told him that she hated him. She repeated it three times before she slammed the phone. He hasn't talked to her since. He works exclusively as a relief pharmacist and told me that the places where he works know exactly what they are getting. A bearded biker pharmacist who tolerates nothing. As he said it, "I take no shit."

I asked him, "So pharmacy for you now is just about the money?"

There was a pause and I remembered him as I saw him twenty years ago. Still handsome with a rugged look with the beard and hair. His eyes were bright and he looked like life was being good. His voice sounded tired. I wondered how he looked now. He said, "Jamie, me boy, it has always been just about the money."

My friend was never a common pharmacist. He was given special privileges because he was a Superstar in the company and they had him pegged for middle management one day. He was never the Prisoner who would allow the Jailer to abuse him. His Jailers learned to just leave him alone. However, he made the same sacrifices that you and I have made. He suffered under the same poor working conditions. I don't know what really caused him to bolt, but my guess is that there was no "Why" good enough to make the "How" worthwhile.

There is a woman in East Cleveland with a "Why" so good that she can put up with just about any "How". Her job is her life, but it not about the money and it never was. She is dedicated to serving a clientele that is very poor. Most of them are on welfare. They are not very well educated, but this fit, blonde white woman is their angel. The Jailers are good to her. They never bother her at all and give her anything she asks for. The Wardens of her company have given The Jailers their marching orders. This woman provides the only stability in the entire store. She has no partner. The hours she is off are covered by floaters. I have a great deal of respect for her. Her job is not to sell drugs, but to serve the patients. I have more respect for her because she takes such good care of herself. The company is pleased to give her the paid three vacations of three weeks every year that she negotiated for. She likes Cancun and the islands of Hawaii. She spent one vacation on a train, traveling across Canada.

The real difference between these two pharmacists is that one was a Prisoner with no "Why" and the other is able to thrive doing a difficult "How" because she has a worthwhile "Why". The woman in Cleveland has a steady boyfriend. He is a doctor who volunteers at the free clinic where they met.

I don't believe for a minute that you have to have a "Why" as heroic as hers is. It is her choice. She likes her life. We all have choices every day. My "Why" is much simpler and it works for me.

We go through three phases in our job of working as a pharmacist. They are distinct and, depending where you are in your career, you will recognize these important chapters.

Young pharmacists do not know that they are Prisoners when they enter the profession. They are all Doctors now. They have been clinically trained. They suffer from a psychological condition called a delusion of hope. They see how older pharmacists are treated. They watch their preceptors trudge through twelve hour shifts. They hear the staff pharmacist arguing on the telephone with their spouse. They can't make it to the school event again and it is not their fault. They see that the pharmacist is either too skinny or too fat from either avoiding bad food or succumbing to candy bars and caffeinated beverages with a yellow mouth from that giant bag of Cheetos. The youngsters will not hesitate to make eye contact and tell you it is not going to happen to them. It will be different where they work.

They will also tell you that they will practice pharmacy as a professional and not just run the Prescription Mill. They know exactly what they are supposed to be doing and they, by god, are going to do it right.

They have high expectations and have those prospects dashed right out of the gate. The Jailers assert their brand of control as soon as they can. The "Why" of a young pharmacist is simply to practice pharmacy in the manner in which they were trained. This is not allowed because of the demands that the Prescription Mill places on their time and energy. They deny it in the beginning, but the pressure of a waiting and impatient public get to them the first week. When the demands of a ringing telephone, people waiting at the counter, a drive-through bell ringing, doctor calls on the voice mail, twelve prescriptions still not filled and the manager commanding that they help the woman wanting vitamins defeat them and they compromise their integrity the first time, the flood gates open and they will drop counseling and professionalism more often as their career continues. They become everything they made fun of while in pharmacy school. They are a deep well of drug therapy information and they use none of it. They hate themselves.

I worked with a young woman in Washington State when she was an intern. She told me on numerous occasions that she knew that she would have to own her own store if she wanted to be happy working in pharmacy. She had watched the pharmacists she had worked with dance The Prisoners' dance and decided that she would have none of it.

Young people are, by nature, idealists. They are going to change the world. They are not going to compromise. They have been thoroughly indoctrinated by their professors that they are important members of the medical community and that their clinical skills will be utilized every day. That is the way it should be. It would be unconscionable if a teacher told them that they will be chained to The Prescription Mill eight hours to fourteen hours straight, every single day.

There are many students who need to work and the well of jobs is the deepest in retail. Most of them work for one of the chains. They watch the veteran pharmacists and come to one of two possible conclusions. They say, "I'm better than John and Mildred." They have hope. "I am not going to put up with what they tolerate. I am going to practice pharmacy the way I have been taught." Or they say, "This is impossible." For these young people it is hopeless before they even start. "Look at John and Mildred. I could not tolerate being like them."

I believe that most new pharmacists enter retail believing that they are different. They keep the delusion that they will not succumb to the repressive pressures of The Prison right up until they can't anymore. They are young and very strong and they have not been pounded on yet so they walk tall for awhile until they start compromising the three pillars of a professional person on a regular basis. I am referring to personal standards, ethical principles and professional responsibilities.

Of these three, it is especially damaging to one's psyche when personal standards are thrown under the bus in order to do the "How" that is required to run The Prescription Mill. I hate it when I talk to a young pharmacist and see that there is no spark in their eyes anymore. There is no more life there than there is in the eyes of a piece-work laborer in a factory. It isn't because something has been taken away from them. It is because they have given away something that was personally very important.

I knew a young woman in Vermont who always said that she wanted to specialize in women's health. She didn't want to attend to women's plumbing. That is well taken care of. She wanted to make a difference with issues like hot flashes, short term memory problems, depression, a sense of not belonging, loss of libido, sexual pain and difficulties, the lack of pleasure in everyday life. There came a point when she realized that she was not doing what she wanted to do and she began to express hatred at her job. You could see that she had given up. She had missed so many chances to help women that she believed that she had failed and that there was no way back. She rarely looked up from The Prescription Mill. You could see that it was easier that way. She was tired and the way back to her personal standards was just too hard.

Ethical principles are debatable. There is grey area between what I think my pharmacist's ethics require and what yours require. My ethics require me to give my attention to anyone who I think is in need. Having such stringent rules for my own behavior puts me in a difficult situation. How can I possibly help everyone who asks for my help? I have learned to discern the difference between people who just want my help and those who are, indeed, in need. The person who argues with me about the advice I have just given him is not in need. I know the difference. I usually ask that person why they are even bothering to ask me if they already know the answer. I also will not hesitate to tell them that the advice that I have given them comes from a font of education learned in school, CE and from many years of experience. If they do not want to take my advice, I have other things I need to do.

Most young pharmacists have gone without money for at least six years. The money they earn in retail can sustain them for a long time. It is easy to look away from ethical principles, put their nose to The Mill and collect that big check. But not without a personal reprimand. We can't escape from ourselves. We are always there and we know when we have failed ourselves. There will be reminders and none of us can look the other way forever.

There is no debate about professional responsibilities. The argument, I can't counsel because of The Prescription Mill Timers is not acceptable. To allow an Intern to conclude that they can break the law with impunity is unacceptable. Interns are too often mentored by pharmacists who have no business acting as Preceptors. Interns and young pharmacists watch veteran pharmacists flaunt the law every single day. None of us would ever consider allowing the pharmacy to be open when there is no pharmacist on duty. That is an administrative issue. That is breaking the law. Patient care is not involved. Neglecting to counsel is neglecting to provide patient care that is mandated by law. The young pharmacist's responsibility to counsel appropriately is a legal liability and they very well may be held accountable if they consistently break the law. There will be some highly publicized busts. In Washington State, Florida or Texas. What happens then? It will be a game-changer.

The three pillars should be enough, but there is much more that diminishes the young pharmacist's life force. The Interns are not pharmacists yet. They can leave the store and go to the local sandwich shop for lunch. They are free to enjoy the holiday breaks at their family's home two hundred miles from the school. They go to the bathroom when they have to go. The idea of a fourteen hour shift with no breaks is not part of their paradigms. Then, they become a Registered Pharmacist, take a job, and on the second Tuesday find themselves still at work after twelve hours with too much caffeine in their body and an empty pound bag of Butterfingers on the counter. The cashier is gone for the day and the Technician has a dozen prescriptions to type. The young pharmacist tries to answer the telephone according to company policy and finds out that answering the telephone is all she has done for fifteen minutes. Never again, she promises, but she decides to forego counseling two hours of patients because of The Prescription Mill Timers. She confronts the Pharmacy Manager the next day. She complains that she is not practicing pharmacy the way she wants to. He shames himself by laughing at her and asking, "Do you think that Cost-Less Drugs cares what you want, Beverly?" The young pharmacist who does not document every single incident like this is very unwise. They do not understand that the law is the law. They cannot ignore it and still maintain their integrity. The real questions is fundamental, one of context: Does Beverly work for Cost-Less Drugs? Or is Cost-Less Drugs the place where she has brought her practice of pharmacy for a price?

Dignity is something that you give to yourself. The Prisoners in The Prison have to fight to be dignified. The Jailers are relatively intelligent people. They would not be store managers if they weren't capable. They are smart enough to know how to get a young pharmacist's goat. They can spot a young pharmacist who is not complying as a Prisoner should and are likely to make getting the pharmacist in line a priority.

Some Jailers will play the company policy card at every turn. Some gleefully rub their hands together when they can play the "Rude" card. All of them will play the you-didn't-get-that-report-in-on-time card. They don't have to play the customer service card because that one is always on the table. The "write-up" is the club they use. The young pharmacist who can hold up against this withering attack by The Jailers is rare. They don't want to take a chance on losing their job, after all. Suddenly, they are in the same boat as the Prisoner with a family, mortgage, two car payments and kids in good schools. They can't afford to lose their job. They are Prisoners. The pleasure of working is gone. The difficulties they put up with are too huge. The "How" is too hard when the only "Why" is money.

If that isn't enough, you have Jailers who are very talented and creative and intrepid. They are like teenagers on dope. They come to believe that they are bullet-proof and they take chances that are not chances at all because the pharmacist is not going to complain to the people who can do something about it. Often, that is the government. The E.E.O.C. would be all over The Jailer who engages in sexual harassment, but is running his fingers through your hair actually sexual harassment? Is leering at your bottom and referring to you as "Lover

Buns" when he is talking about you to the Assistant Jailer actually harassment? These are harassment and, if they put up with them, the young pharmacists are willingly giving away their dignity because they are afraid of The Jailers. They are given a club and they don't use it.

Young pharmacists do not enter the profession expecting the worst. The come in with rosy expectations of practicing their chosen profession the way they know it is supposed to be practiced. They soon realize that the divide between the job and the profession is huge. There is a slow erosion of their best expectations. They give up and just survive as Prisoners. Their expectations are shattered. What are they supposed to do when swimming out in the middle of the stream where the water runs clear is just too hard. They see everyone else wallowing in the muck. What do you expect them to do? Stand up to the Jailer all by themselves?

The veteran pharmacist who has given up and just can't stand the enthusiasm of youth should be censured. I watched a Pharmacy Manager tell a young pharmacist that he would be smart to just "Get with the program". He told the young man that his life would be better if he simply came to work every day expecting the job to be a "Piece of Shit". He suggested that "Making waves" would make the young man's life miserable. This is abject cruelty. The young man did exactly as he was told. The excitement and purpose in his life was robbed from him and he became a Prisoner himself.

I have read complaints from hundreds of pharmacists. They blame the profession. I have heard from pharmacists who claim that they will do everything they can to keep their own children from choosing pharmacy as a course of study. It only takes a few years and some Prisoners start looking for a way to break out of Prison. There are teachers and real estate agents who are pharmacists. Drug Company representatives are pharmacists. There are stock brokers who are pharmacists. There are also plenty who left to build a career elsewhere and came back. The money is just too good.

Young pharmacists see all of these scenarios and, justifiably, ask themselves, "What did I get myself into?" They no longer think that they are different. They do not believe that are too strong to succumb. I have heard from scores of young pharmacists who claim that they will make big changes just as soon as they have fulfilled the conditions of their sign-on agreement.

A number of years ago, one of The Governors (Executives) of one of the major drug store chains was the featured speaker at a meeting of some of the companies pharmacists. They were all Pharmacy Managers. This Governor apparently wanted to show off and exhibit how well versed he was in Psychology 101. He was a Governor, after all, and a Governor's ego is often in the Masters of the Universe class. My friend in Seattle, the artist who wants nothing that smells of pharmacy in her life away from work, was in the room. This talk put this company behind by a decade in recruiting the best pharmacists. It is no wonder that my artist friend quit her job and emotionally bailed out. The Governor addressed Abraham Maslow's

Hierarchy of Needs.

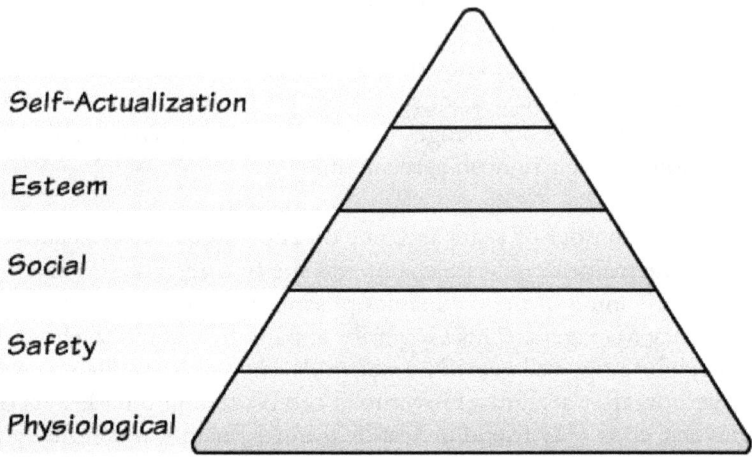

Maslow's Hierarchy

Self-Actualization

Esteem

Social

Safety

Physiological

He started at the bottom and told the assembled Prisoners that their job made all of the basic biological and physiological needs at the bottom of the pyramid possible. The company paid enough money that the pharmacists could provide all of the food, drink, clothing, shelter, a place to sleep and warmth that they and their family needed. My friend reported that the executive made a leering crude comment about the pharmacists' sexual lives. Our artist friend said that he spoke in a loud, insistent manner as if he was daring someone to argue with him. He was a Governor, however, and even though there was a lot of uncomfortable shifting in the seats no one said a word.

The next level up is where the Safety Needs are listed. The executive insisted that the company paid enough money that the Prisoners could afford to live in a law and order community where there would be good police services to provide safety and security. He said that the pharmacists made enough to provide stability for their families.

He said that it wasn't the job of either pharmacy or the company to see to it that the pharmacists felt like they belonged to anything. He told them to join a church. My friend said that, at this point, The Wardens in the room seemed uncomfortable.

He spent an inordinate amount of time on the levels from Esteem Needs all the way to the top. He told these pharmacists that it was ridiculous for them to think that they could get self-esteem or self-respect from their jobs. That is not what a job is for. A job is to make money to provide for the bottom two levels.

This was the beginning of the end for our artist friend. She works part time now. She refers to it as chopping wood and carrying water. It is just about the money. She hates doing it, but the money is good for carrying water. She has beauty, balance and form in her life now. Her achievements are artistic and she has a reputation as an artist. There is personal growth and she does help others find their way.

How do you tell a twenty five year old that his professional life is not going to help him find knowledge and self-awareness? Too many pharmacy managers believe exactly what the Governor in our example believed. The pharmacist is just a worker. That is not what professionals believe about themselves. This is a toxic stew fed to young people who are hungry for meaning in their lives and want to know the way.

Thus, for the young pharmacist, begins the barren landscape of a career stripped of creativity, commitment and noble purpose. There is not good enough "Why" to make up for a "How" when the Prison feeds unacceptable-to-any-other-profession behavior like, "Hey, you, where is da lawn chairs for nine ninety-nine?" or "You're just a pharmist. You can't tell me that these refills are no good just because it is six months". Just the proximity of the work area to the customers is demeaning. They put you right out front where anybody can get to you. Of course, the pharmacist is going to look at the young attorney seeing clients by appointment and charging in ten minute increments and think, What the…?

Dostoyevsky said: A man can get used to anything, but don't ask us how. It takes no time at all to become entrenched in Prison life. There are things that happen that can cause a pharmacist to almost lose all reason or they have none to lose. Thus, the illusions of an ideal life as a pharmacist are destroyed one by one. Most of us are overpowered by a grim sense of humor. It is necessary for our survival. There is nothing to lose anymore when we have so easily given away our self-respect and integrity for the money.

Have you ever listened to a group of pharmacists when they have had a few drinks at a social function? They fall into laughing about their situations. They play You think that is bad, listen to this. Then they tell their story, trying their damndest to make it worse than the other guy's story.

They laugh as they tell about working all alone after a certain time in the evening. The clerk leaves at six and the last technician at seven and there are two more hours to go. It is raining and cars are lined up at the drive-through. The third in line starts honking and number two gets out of the car. He is a big man and he walks back to number three and stabs his finger at the closed window. Next thing you know, he starts pounding on the driver's side door of number three. There is no more honking after that. The telephone is non-stop ringing, but it has to be ignored. Answer the phone and you know that you are dead. They never give you a straight message. They always have a story they want to tell. Two lines are ringing now.

There is an irate woman at the counter. She tells you how long she has been waiting and you haven't even begun to work on her three prescriptions. A nice elderly man holds his hand up, "Take your time," he says. He hands you four prescriptions. You see that he has neatly printed his name, address, phone number and date of birth on each one. You want to kiss him. The floor manager makes an appearance. "Why are these people having to wait? They just want to buy their prescriptions." He huffs and puffs. He looks a little intimidated. He is a Jailer-in-Training. "Why are you just standing here? How long has the drive-through been waiting."

This is a modern pharmacy. White and tile, but it is now pure ugliness. The pharmacist experiences disgust and failure and hopelessness. The pharmacist's feelings, ambitions, standards, ethics and legal responsibilities have all been blunted in a matter of minutes. Walking out is not an option. The mortgage must be paid. The kids' schools. The club membership is new. The spouse makes a fraction of what the pharmacist earns. This means there is no freedom. The fight or flight instinct is smothered and the stomach hyper-acidity is worse than ever. The pharmacist gives up and survives until nine o'clock. Why is this funny? I almost expect them to break into song, like it was picking cotton.

The pharmacist is numb on the drive home. He is trapped. His throat is filled with the burning of the acid of disgust, horror and self-pity. He brushes past his spouse and ignores her greeting. He sees warmed up meat loaf on the table and barks at her, "I want a pizza. Order a damned pizza. How many times do you think I can eat meat loaf?" He takes a beer from the fridge and goes to his room. The game is still on. He kicks off his shoes, sits for awhile and then gets up for another beer. He is asleep when the pizza delivery arrives.

Two days later, he comes home at nine thirty and the house is dark. Inside, he finds that his wife and kids are gone. The closets are empty. Most of the furniture is gone. There isn't even a note. The pharmacist ends up falling to the floor. He is wracked with sobs. He rubs the mucus away from his nose. He is devastated at what he has caused. But, at the same time, there is a wave of relief that surrounds him. He can't do it with them, but he may be able to do this all alone. He can support them like he is supposed to. All he needs is a small apartment. He knows that he will be okay. Then, he realizes that he has to tell his mother and begins crying again.

That may be dramatic, but our pharmacist survives. The conditions of working in the pharmacy become so commonplace that after a year or two he is no longer moved. He has no problem ignoring the telephone when they take his help away and he is stuck working alone. He knows how to survive. If he chooses to keep the drive-through open, the drive-through customers wait just as long as the people at the counter. If they honk, they wait longer. He takes care of each patient in turn. The waiting time is two hours. He gives each prescription the same care he would give for an order for his ten year old. The floor Manager keeps away. This pharmacist has a reputation. The word is that he can be mean. When there were numerous complaints about his service, the District Manager came to the store. The store manager was with him. They stood aside, expecting the pharmacist to react, but he just kept working.

The Warden cleared his throat. "What seems to be the problem?" the Warden asked. "People complain every time you work the late shift that the wait time is excessive. You have to do better. You know the company policy on wait times."

The Jailer pipes in, "You don't answer the phone. You let it ring." The store manager is emboldened because the pharmacist district manager is doing the heavy lifting of reprimanding.

"Give me back my clerk and the technician and the wait times will be what you want."

"You know that we can't do that. There is a budget and we will intend to stick to it." The Warden stares at the pharmacist. "Everyone has to suck it up."

"I have no choice but to write you up." The Jailer is smiling.

The pharmacist has a new image. He wears a goatee. His hair is longish. He lives alone and his only stress relief is working out. He takes one step toward The Jailer and the puffy store manager retreats two steps. The pharmacist looks at the manager and smiles. "Fuck you and fuck your write-up, man."

The Warden stiffens. "You can't talk to him like that. I will write you up and you will sign it."

"Fuck you too, man." There is a pause. The pharmacist's eyes harden. "Fire me." He holds his hand up to silence The Jailer. "Do you know how many times you have told me that I counsel too much? How many times have you ordered me to speed things up by not counseling at all? I can tell you the exact words you used because I have documented every single incident ordering me to break the law."

The Jailer huffs, "I never... "

"Don't," The Warden stops The Jailer. "Don't say another word."

I know that this little fiction is extreme, but is it really that much over the top? It is a story that is at least partially true every single day. It is pathetic, but pharmacists don't have a say in how they practice their own profession. They just follow the rules. It does not matter that they are highly educated medical professionals. This is not right. They are Prisoners first. Dignity? What a laugh!

There are pharmacists who work an entire career like this. The conditions become so commonplace that the pharmacist is no longer moved by anything. The paycheck at the end of the week is all that matters. It is the only "Why" that counts. The pity is that The Prisoners are so driven by the rules that they have no problem throwing each other under the bus.

Another interesting sensation that occupies us is curiosity. Perhaps it is beneficial for our mental health. We actually have the ability to be observers of our situation. We are curious if we are going to come out of this okay. We are especially interested in how other pharmacists get along in their jobs and we want to know what is going to happen next to us. What is going to happen when the store manager finds out that the pharmacy is over budget in labor hours? Will there be trouble when the inventory shows how overstocked we are? Will they give me my vacation in June? The family reunion comes only once every decade. It is my husband's family and he says that he'll take the kids and go to Iowa without me. How will that look? His aunts never did like me very much.

We find out that everything we held sacred is a lie. The state boards of pharmacy don't care that we are dangerous when we are hungry and tired. Pharmacy boards are mandated to protect the public from dangerous pharmacists and they don't recognize that every pharmacist in the twelfth hour of a fourteen hour day is a dangerous pharmacist. Can't they see what is happening? Pharmacists write to them every day and complain and ask the help of the boards. They answer that it is not their job to regulate working conditions, but can't they see that we are actually commenting on protecting the public?

Pharmacists make life or death choices all day long. After too many fourteen hour days in a row, as one company requires, the pharmacist is dangerous every single minute. Have you noticed how difficult it is to sleep after a long day without a rest period or a regular meal break? Some of us fill prescriptions all night long. My habit was to fill the same prescription, over and over, and never get it right. You go home at ten o'clock and go right back to open at eight o'clock. Is it any wonder that we make mistakes? It is a miracle that more people don't die.

In the end, we all are curious about each other and end up doing what we do every single day, many of us hating every minute of it. We diminish the seriousness of our dilemma. We laugh about it because if we didn't laugh, we would cry.

My last management job ended around 1996. A couple years before that, a middle-aged pharmacist who had been out of work stopped by the pharmacy to talk with me. He wanted to work part time. I told him that I had a regular part time pharmacist and that Mildred seemed to like the job and had not indicated that she wanted to leave. Her commute was an hour one way. She was always on time and she had never missed a shift. She was loyal and I would be loyal to her.

"Is she having trouble doing the job?" this man asked me. He gave me a knowing look that made me uncomfortable, like the look one might give you if he was trying to tell you about a wayward spouse.

"She is doing fine," I said. "Mildred is a good pharmacist." She was slow, but in 1996, the abject Prescription Mill craziness that we put up with now had not yet begun. There had been complaints, but not about her competence. The complaints were always about the wait times late in the day when she worked alone.

"Dave told me that you might be thinking of a change." That look again. What did he know? What was he trying to tell me.

"Dave told you?" Dave was store manager. "What did Dave say?"

"He just said that Mildred was having problems."

"Dave should mind his own business." This was coming from left field. I suddenly did not like this guy as much.

Then, the zinger. "You knew that Mildred has a drinking problem didn't you?" That look again. This time with eyes arched. This pharmacist was throwing Mildred under the bus, hoping to snag her two day a week job.

"She hasn't looked impaired to me,' I said.

"Have you ever smelled her breath?"

I kicked the guy out. I told him not to come back. The next day, Dave, The Jailer, jumped me. "Mildred is a drunk," he said. He demanded that I fire her and hire the hard core Prisoner who was not averse to using gossip for his advantage. He had tried to throw Mildred under the bus.

Of course, I refused, but I did have a talk with Mildred. I asked her if she had a drinking problem.

She released a long sigh and looked at me. "It never ends, Jim. That was in a previous life, during my first marriage. I've been sober for twelve years and four months. Who told you?"

I had no problem telling her who the man was. "He will never work here," I added, "Not while I am the pharmacy manager."

Mildred nodded. She thanked me. She stood tall and showed a self-respect and dignity that is often lacking in that environment.

Apathy is a way of life in the retail chain pharmacy world. Apathy is not caring and not caring that you don't care. The pharmacist doesn't think they can do anything about it anyway so it is better to not care. That makes it very easy to look away when our fellow pharmacists have problems. It also makes us very alone when we have our own problems.

There was an older man who had worked for a chain drug store company for thirty years. He was a competent pharmacist who had graduated from the typewriter and calculating price with a pencil and notepad era to the calculator to the computer. He was not comfortable with the computer and the computer in the 1980s was not user friendly. He took an inordinate amount of time to process new prescriptions. His problem was that he thought he could hurt something by not being perfect. The store manager fired him. He called him to the office and showed him pages and pages of complaints about his being slow. Customers said that he was grumpy and short with them. A doctor complained that this pharmacist refused to make a compound on a day when he had no pharmacist overlap. The store manager was up to something and everyone who worked in the pharmacy knew it. He had hired a new pharmacist before he even fired the older man.

The new hire was a pharmacist whom the manager had known for a few years. The Prisoner knew that he had given good service to the company for thirty years. He came to the two other pharmacists to enlist their help. He was desperate. The drug store was on Whidbey Island and the closest possible job was on the mainland, an hour away.

The other pharmacists shrugged and looked away when he talked to them. They changed the subject. They gave him vapid advice about where the jobs were. They were not fond of the store manager and they hated what he was doing, but they believed that they were helpless. They simply knew that there was nothing they could do to help our friend. They did not want to get on the bad side of the store manager, so they pretended as if everything was okay. They had no emotions about the firing. They were dulled and disinterested as if it couldn't happen to them. They knew that it was a sucker's game to care about anything anymore.

Pharmacists learn that they can endure just about anything as long as the paychecks keep coming. I knew a woman whose husband was mentally ill. He was schizophrenic and he was not much of a husband. This pharmacy manager and the divorced staff pharmacist began an affair one night when they were both tired and lonely. They kept this up for a year or two. The staff pharmacist was satisfied with the respite from his solitary life, but the pharmacy manager was consumed with guilt. Her husband knew about the affair, but that didn't help.

She was a woman who was raised in the south in the 1950s. When you married it was forever and you were always faithful. This woman's guilt was palpable on the days after the night before in a bed at an out of town hotel. She suffered greatly and demanded of herself that it never happen again. That was repentance and she was good until a few weeks later when it happened again. Her lover did not deserve such a loyal woman. He kissed and told and the pharmacy manager endured great emotional pain just from the knowing looks and the snide remarks. She endured though. Eventually, her lover became bored and moved on. It was for the best, she knew, but she still had a schizophrenic, house-bound husband and she cried at the loss of the diversion.

Would this scenario be different if she had been happy with her job? I don't know, but I'd guess that she would have had a better chance at managing her life in a manner that brought her some self-respect. She had not acted with integrity and she knew it. She castigated herself as no one else could.

The Prison has its own unique form of punishment. It is called the Write Up. The Jailer uses the write up when a customer complains about you. He will write you up when come in late to work. He will write you up if you do not follow company policy. I am certain that a company not named Three Pee Ex authorizes their Jailers to write up a pharmacist for not toeing the mark with the timers on the Prescription Mill. Conceivably, the store manager has the latitude to write up a pharmacist for just about anything he wants. That is a formidable club to put in the hands of a person who doesn't like you and may even hate you because of the money you make and your importance to the company.

I worked with a young woman who was a wonder. She was a Pharm D in the mid 1980s and she insisted that her nametag read Doctor. She worked her twelve hour shifts wearing a skirt and blouse and high heel shoes. She was energetic, beautiful and very smart. The Jailer hated her. He hated her because she didn't listen to him. He hated her because he wanted to date her and she laughed at him. He hated her because she warned him about sexual harassment when he ran his fingers through her hair one morning. He told her that there was something in her hair and he was just getting it out. The store manager wanted to ruin this woman.

One evening, two attractive young girls brought in a whole stack of prescriptions ten minutes before closing. The pharmacist was ending a twelve hour day. She was tired and none of the prescriptions needed to be started right away. She told the girls that they would be filled the first thing the next morning. The girls went ape shit. They used abusive language and the pharmacist refused to even talk to them. They were continuing the name calling and the use of profanities right up to when the pharmacist locked the door to the pharmacy.

The next morning, the pharmacist on duty thought it was odd when The Jailer asked for the prescriptions. He said that the girls had changed their minds and did not want them filled. The morning pharmacist did not say a word to the pharmacist from the night before. He did not want to get involved. A customer who had witnessed the altercation called the pharmacist at home and reported that the girls were bartenders at a strip club where the manager spent his evenings. It was a set up.

When the manager called the pharmacist to the office for his lecture and write up, the pharmacist accused him of setting her up. He denied it, of course, and then told the pharmacist if she wanted to keep her job, she would have to write letters of apology to the two bartenders. She told him to suck eggs, refused to sign the write up and went home early, thoroughly shaken. What was killing her was not the punishment for something she didn't do. It was not the write up itself. It was the unreasonableness of the entire situation, the injustice of it. To top it off, she had no allies. The other staff pharmacist told her that he did not want to get involved. The Pharmacy Manager refused to send an explanatory note to the district pharmacy manager. The Warden only knew that she had refused to fill the prescriptions. The circumstances were not included in the write up. It was the insult that was the most painful part.

There are rewards for being a good prisoner, one who behaves and follows all of the rules, even if they are wrong. The Jailer and the Warden might make sure that you get the vacation that you want while all of the other pharmacists have to fight for a week in the summer and piecemeal out the other three or four weeks throughout the year. The compliant pharmacist may get the best holidays off and never have to worry about having to work on his children's birthdays. I knew a pharmacist who played golf with the store manager regularly. They were real buddies and the resentment in the pharmacy was like a fog. When the pharmacist wasn't at work, the talk was vicious.

This pharmacist was almost like a trustee in the prison. The Jailer was all over the other two pharmacists about hurrying up, but he never said a word to the trustee and this guy was the slowest of the three. When the company held a summer picnic at a theme park, the trustee was scheduled to work. The manager changed the schedule two weeks before the event so the trustee could attend with his family. It didn't matter that the woman who all of a sudden had to work was a single mother with thirteen year old twins. She complained vociferously that it was not fair and the manager said that he didn't care. "Haven't you noticed?" he said, "Life is not fair." With that, he laughed and dismissed her with a wave of his hand.

There are punishments for being a bad Prisoner. Often, they are arbitrary and they represent a Jailer seeing what he can get away with. If the store manager does not like you and that is often the case, he'll try just about anything, based on a whim. The store manager that disliked me the most was a man in his fifties. He was an old hand and was used to being the lovable fatherly type to the female employees. He eventually was walked out of the store by security personnel for sexual harassment. A female pharmacist was the messenger who called the kettle black. She asked him one day how he expected to continue to get away with sleeping with both the bookkeeper and the cosmetician. Harassment was not what she was talking about. She sincerely wondered how he could be carrying on sexual relationships with two female employees at the same time. She believed that he felt threatened because she dared to voice what everyone wanted to know.

He began a campaign of harassing the pharmacist. One morning, he walked into the pharmacy and stood there silently, watching her work. She asked him what he wanted a few times, but, when he wouldn't answer, she just ignored him. The technician was clearly nervous. The pharmacist told her to just do her job and that to ignore the manager because he was harmless. I think that all she could see was a man having sex with two of her fellow employees and who could make her life miserable.

At one point, the Jailer said, "I see that you have a hard time keeping up. You have not had the counter clear for over thirty minutes. I'm giving you a half hour to get the counter clean or I am writing you up."

The technician took it personally and argued, "But they keep bringing in new prescriptions."

"I'm not writing you up, I'm going to write her up."

The pharmacist made the mistake of laughing. "You are writing me up for what?"

"For inferior customer service. You are making the customers wait unnecessarily." He was sputtering.

"Nobody has had to wait longer than fifteen minutes." She went back to work.

"That is too long. Company policy is no longer than ten."

She was strong. She was not intimidated, but those years when she worked with this store manager was when she started taking ranitidine. These kinds of mental conflicts are soul-destroying. The clashes of will-power that we have to tolerate are impossible for other professional people to believe.

This Jailer judged the pharmacists every single day, but his knowledge of pharmacy and what pharmacists do was limited to the rules of the company. He did not care that it was the pharmacist's job to make sure that a young mother was educated on how her four year old was supposed to use an albuterol MDI. He did not care that compounding takes time and that the average chain pharmacy does not have all of the supplies and equipment needed to do some compounds. He did not care that pharmacists have the right and responsibility to refuse to fill prescriptions when they are for #360 Norco-10, #180 Xanax 2 mg and #180 Somas 350 mg and the patient has driven sixty miles to try to find a pharmacy who will fill them.

"Well, the doctor wrote them. They are legal prescriptions. Fill them." He didn't want to hear about doubts that the drugs were not for a legitimate medical condition. To the credit of all of the pharmacists, they dug in their heels and refused to be intimidated. This store manager's ignorance about pharmacy was astounding and he was still The Jailer. To be judged by someone lower who had no idea of what a pharmacist's responsibilities are is upsetting to some pharmacists and depressing to others. I know pharmacists who take their dose of SSRI every single morning simply because they feel that they do not have control of their own lives. They feel helpless to change their circumstances. The future looks hopeless to them. They know that a professional does not wet their underpants, but it has happened to them and the Jailers complain loudly when the Prescription Mill comes to a halt while they are in the bathroom.

Every single pharmacist knows that lawyers take a lunch whenever they want. Attorneys charge for everything. Doctors are treated with consummate respect and deference and the pharmacist is referred to as "Hey You" and the questions are often in the category of "Where are the flip-flops on sale for one ninety nine cents?"

Pharmacists are treated badly by The Jailers, the drug store customers, legitimate patients and each others. The Wardens know what is going on in The Prison, but they only provide lip service and, with their silence, condone the poor working conditions and the lack of respect for pharmacists. We all know that what we put up with is appalling. If you are unlucky enough to work for a non-traditional drug store company, it could be downright toxic. The Jailers in the big box stores and grocery stores know nothing about the pharmacy business and they could not care less about what you put up with. For non-drug store companies with only a limited number of stores, it is even worse. There are no strong pharmacist Wardens. The pharmacist is a disposable employee. They don't need you to make a profit. Some of them don't even want the pharmacy department in the store. The inventory is too high and your wage downright angers the Jailers. They diminish what you do and, frankly, treat you like an interloper, an unwanted indentured servant, if you will.

To make it worse, the high minded professional tasks in the Medication Therapy Management category are not available professional jobs for most pharmacists. When you do administer vaccine or manage diabetic patients, they do not make sure that you have the time and all of the important tools. They certainly do not shutdown the Prescription Mill while you are with a patient in the counseling area. They don't pay you for it either.

I have never personally been so downtrodden and had so little confidence in the job of working as a pharmacist that I wanted out. But, I am an optimist. Of course, I have felt hopeless at times and I have been stupid enough to think that I can change things. Most chain pharmacists feel the same emotion at the end of the day. Relief that the day is over. They go home and, if they are polite, they tell the spouse to just leave them alone for an hour or so. Some are so angry and beaten down that they abuse their spouse.

We are often treated like just another employee by The Jailers, but pharmacists are the only employees in the entire store who are expected to work with zero errors. Prescription errors could conceivable cause great harm to the patient, but the store managers are more likely to come down hard on a pharmacist who makes ordering errors and is overstocked. The error of being over the labor budget is an even more egregious error. They practically do not even pay attention when there is a misfill, but can get irate when the pharmacy manager inadvertently schedules a technician for a holiday that ends up in an overtime situation. The priorities and the values are totally misapplied. And the pharmacist bites and suffers more when a technician gets a few hours of overtime than they do when they let a prescription for lisinopril get filled for lansoprazole.

I worked with a pharmacy manager who dispensed a non-drowsy antihistamine instead of nifedipine. This was in the 1980s and prescription volumes were low enough that the pharmacist often worked without a technician. The error was all his. A few weeks later, the patient, short of breath and pallid, asked me if the drug was correct. I told him the truth. The store manager did not seem to care that much about the error, but he was zealous about the write up. The pharmacy manager was more concerned about the black mark that would go in his record than he was about possibly causing damage to the patient.

I can't imagine any other profession that can cause as much cortisol to stream through the veins than pharmacy. Zero defects is an unreasonable expectation. However, we deal in a commodity and the evidence of an error is apparent. When a doctor makes an error in diagnosis, nobody has to know. When an attorney interprets the law wrong, it is often ambiguous and open for debate. Not us. If you dispense Seldane and the prescription was for Procardia, there is no ambiguity. There is no grey area. You are wrong. The Prison will punish you and, pathetically, they will punish you not because of the error, but because it is a chance to get back at you for some misty wrong that you have committed. Often, it can be because you make more money that The Jailer. Now, that is a crime.

There is so much riding on what we do or do not do and the professional choices we make that the citizens of the states have every right to expect that the state boards of pharmacy would make sure that pharmacists work under conditions that will allow them to be both physically rested and mentally sharp at all times. That is a joke that only pharmacists can really appreciate. It is no wonder that so many mid-career pharmacists still dream of getting out. They are Prisoners of Comfort, however, and most would not or cannot take a cut in pay.

When I was a very young pharmacist, well before there were computers and such a thing as a Prescription Mill with timers and reports on wait times, I hated it that I had to go to work five days a week and that I couldn't just up and go anytime I wanted. It did not take me much time to get over it because I knew that I liked the living I was making. I lasted almost ten years before I took some time off. Lots of time off. From November, 1976 until May, 1981 I did not work more than twenty hours in any one week.

There was a stretch of almost eight months when I did not work at all. I was so tired of doing what the establishments expected me to do. Pharmacist friends hated what I was doing because they were too trapped by comfort and could not even imagine quitting their job and taking off. My mother and father and brother didn't like how I was living, but it was not their life.

My car was paid off, however. I did not have a mortgage. I had a year's wage in the bank account. My second wife was a young woman ten years my junior who had lived a hippie life and thought that what we had was pretty neat. If you want to escape the hum drum pharmacist's life, there is a trade-off. You don't have to work as a full time pharmacist to have your dreams come true. They just have to be a different brand of dreams. There are not that many jobs that pay what you are making.

You learn in grade school about the importance of being properly nourished and eating good, healthy food. You are a health professional and you counsel people on the value of a balanced diet, not too rich in useless calories or excessive amount of fat and sodium. We could hold debates among pharmacists about the demeaning message that a diet of Butterfingers, salty snacks, coffee and Diet Cokes tells you about what a pathetic life you have on the job. It is pitiable and we are wretched actors if we believe we have to put up with it. The store manager eats a nice lunch, unhurried and relaxed. You can count on that.

The most ghastly moment of some pharmacist's day is the first. Unlocking the pharmacy door and hearing the telephone already ringing causes some pharmacists' hearts to sink. When they see that all three lines are lit up, they are defeated before they even start. The technician is not scheduled until nine thirty and the pharmacist is expected to navigate the first half hour all alone. This is professional undernourishment. It is soul-destroying. This is when a pharmacist starts taking real estate broker classes in his spare time. Not because they have always wanted to show houses, but because they feel desperate.

A young woman pharmacist knew when she was in the parking lot if she was defeated. If the store manager's car was not parked, then the day looked okay. She started hyper-ventilating before she got in the store if the store manager was on duty. He didn't do anything but leer at her as she walked into the store. He was always waiting for her in the lobby of the store and he always undressed her with his eyes. He never did anything overt, but his covert harassment caused her to quit a job that she had ten years invested in. The pharmacy district manager tried to talk her out of quitting, but she would not back down. She was too ashamed to tell him the real reason. For some reason, she thought that the harassment was her fault.

What was most troubling was the attitude of the people this pharmacist worked with. They acted like survivors. They were relieved that they were not the ones that had to leave a good job. The pity was that they were so thoroughly institutionalized that they thought that what had happened was acceptable and normal. There was a complete lack of sentiment or empathy. If they cared at all about the pharmacist, they did not show it. This manager had every single employee isolated. They were not capable of standing together for anything. This is the condition that pharmacists often find themselves. On an island, all by themselves with nowhere to turn for support.

Then, a few months later, a new employee was very well aware that what was going on in the store was not normal. She made some phone calls and the pharmacist was invited to one of the other female employees home for a meeting. In attendance was a woman from the personnel department and two members of the Loss Prevention office. Two days later, the store manager was escorted from the store. It was too late for our pharmacist. The store manager had been with the company for many years. He got out with retirement. The company paid him off with a handsome severance package.

I have noticed something that interests me. No matter how bad the pharmacist thinks it is, they almost always get through it okay. Perhaps it is because of the money they make. I can't remember any pharmacist who fell through the cracks completely. Money is the great equalizer, remember? Money buys diversions and the smart rat whose job is a piece of shit, in their view, can always afford to go find fulfillment somewhere else.

I remember a pharmacist who traveled from bridge tournament to bridge tournament. He was a cerebral tiny man who did not have the physical tools to play men's games like golf or basketball. He did not like chess, although he was very good at it. He said that chess players were, for the most part, surly and unfriendly while the people who played bridge were easy to get along with. He lived in a small town where they played pinochle so he had to travel and he loved that aspect of the game. He even met a woman who became his girlfriend. It was a miracle because he was five feet two and she was a good four inches shorter than he was. One evening, I was teasing them about having to sit at pillows at the bridge tables. He laughed, but she just smiled and pointed her finger at me. "Don't tell anyone," she said, "I do sit on a pillow." What twinkling eyes. "It's a secret." This guy was perfectly happy with his job. He lived to play bridge. Like our artist in Seattle, pharmacy was simply something he had to do. It was just chopping wood and carrying water. This man had not found a way to simply survive. He thrived. Not all of us are that lucky.

I know a woman in the east who hated her job so much that she wanted to quit and go to work mucking out cages at a zoo. She loved animals, she claimed, and she believed that she would be fulfilled working for a very small wage doing something she thought she would love doing. I asked her how she thought she would like it if it was February and the lion poop was frozen to the cement and it was her job to scrape it up with a chopper and load it into the wheelbarrow and wheel out to the community animal poop pile behind the elephant house. She said that she hadn't thought about it that way and admitted that her fantasy only included spring weather. I asked her if she hated her job so much that she wouldn't be able to tolerate doing it part time and reminded her that a part time pharmacist job would pay more than double an entry level zoo workers job. That was all it took. I don't know how it turned out in the long run, but for a few years she acknowledged my contribution to her happiness. She did go part time. She did make enough money. Her second income was almost as big as her husband's first income. She was a happy camper, working three days a week and still getting benefits and living her life. Her promise to volunteer at the zoo fell by the wayside.

I know that the job can be lemons, but there are plenty of women who have made years worth of lemonade working part time while they had school age children. It is perfectly okay with them that they spend fourteen hours straight at the Prescription Mill. Probably because they only work two days a week, never on weekends or holidays and never on nights when their child is in the school play. I know many male pharmacists who hate their jobs even more because of this. They believe that the part time female pharmacist who is a mother gets preferential treatment. They are right, but, in the era of pharmacist shortages, what do they want? No part time pharmacist? These women are very smart rats and I, for one, say, "Go for it!"

I have heard stories of pharmacists bolting and finding a different way of making a living, but I can't think of one of whom I actually know who has pulled it off. How would you like to be a real estate broker in 2011? A man in Las Vegas sent me an e-mail a year ago. He asked me if a certain big box store was really as bad to work for as a pharmacist as he had heard. I had no idea, but wanted to know more. He had made a lot of money in real estate before the sub-prime fiascoes and after not making one penny in real estate for an entire year his wife got sick of him brooding in front of the television set. She gave him an ultimatum. Go back to pharmacy or get out of the house for good. He went back to pharmacy. He wrote me again and told me that he couldn't remember why he left in the first place. The filter before his eyes had changed dramatically. Come to find out, upon inquiry, the evidence is that this particular big box store is a terrific employer for a pharmacist. He lucked out.

For at least a few of the years when I was pharmacist in a Prison environment, I was virtually absent from much of my life. My marriage had been basically doomed from the start. My wife and I were too dense with needs to notice it.

To make it worse and allow some purulent festering, my situation at work dictated my relative emotional health or exacerbated the lack of such. I would come home from work and tell my spouse that I needed to "unwind" and asked her to leave me alone. I resented it when she told me about problems in her life. Didn't I have enough of my own? I dearly loved my small step-daughter, but I wanted her Mom to handle every aspect of her life. The child's father paid a measly seventy-five dollars a month child support when he paid at all. It was some kind of hippie arrangement made when money was not supposed to matter. I paid for my step-daughter's Montessori School and everything else. That should be enough, I believed. I did not care how much money my wife spent. My life was all about working and escaping from work.

In that marriage, I was absent sexually. I did it, of course, often reluctantly. It was almost like I had a grudge. I would do it with forced vigor and enthusiasm, but it was almost like masturbating with a partner. I just wanted the release and I wanted it hard with squeals and rapid breathing. I wanted it different every time and if there was any hesitation to experimenting, I'd get moody and pissed off. It was two people who had done it so many times that it was pure habit. We merely remembered how it was supposed to be and didn't really experience it at all. We knew how it would begin. We knew how we would do it, how long it should last and how it would end.

After, I would light a cigarette and she would dramatize how disgusting the smoke was and that she still could not believe that she had married a smoker. I'd lay there with a sweaty chest, drawing in huge gulps of smoke and exhaling into the air. I'd say things like, "Divorce me then" and "Kiss my ass." I wanted so much to tell her that I married her to be a father to her daughter, a child I adored, not because I loved her. It was pathetic and disrespectful. Emotionally, I was so consumed with being a Prisoner that I had nothing left for this wife.

I don't know if I can blame my job as a pharmacist for this dreadfulness. I know that it contributed to the misery. I am an idealist and my job was so far from how I thought how I should be treated as a professional. I was disgusted that I valued the money so highly that I allowed it to run my life. By then, I had a mortgage and car payments and two children, however. I had a new car. I enjoyed frequent trips to Bodega Bay, on the Sonoma Coast, or Lake Tahoe with a pocketful of cash for the Blackjack tables. It was a very shallow period of my life, for a very shallow man.

There was a time when my marriage was in limbo and we were separated due to my wife's straying. I took a mistress. You will be interested to know that she was a pharmacy clerk. There were no technicians in the day. She was beautiful and willing and, with her, my job did not interfere with a robust and healthy sexual relationship. She got everything I did not give to my wife. We were like two children finding a new game to play. There were rules and respect. The experimentation that I wanted was welcomed by this woman. This sexual relationship was a refuge and safe haven from the spiritual ghetto that was my authentic life. It didn't last very long she rolled over in bed one afternoon and asked me for a raise. She was a pigette and I was a pig. I fired her the next week.

I firmly believed that since I was tolerating such abuse and disrespect at work that I should get a free pass into the world of the doctor and his nurse, lawyer and his secretary. My life force was depleted by this affair. It was much too intense for a man with some normal sensitivities. I was guilty of being an adulterer even though I was not living with my wife. I was a man who was raised in the 1950s. Had I not engaged in this clandestine relationship, I may not have gone back with my wife and stayed with her for another twenty years.

There are a handful of male pharmacists on the west coast who were my fellows in this club. A pharmacy manager for a major chain confessed that he had a big problem. His job was difficult. He said that he got no respect and hated every minute at work. He had a wife and two teenage daughters. He was a Bishop in his church and that is another story because men who are Bishops do not do what he did. He supposed that he should be having a good life. He blamed pharmacy for all of his problems and said that he should have been a teacher like his mother and father.

He also had a mistress and a third daughter with her. He said that he loved his mistress and the new daughter was the joy of his life, but that he couldn't leave his wife. I asked him why not and he said that it would look too bad and that his two teenagers would hate him forever. In the end, this man, who could not choose, had to live life with no wife and no children. His mistress married another man and he was excommunicated from the church.

I believe that all of us men who looked elsewhere have thinking, feeling, suffering scheming minds and we were looking for that steam control. A willing woman was that release. I suppose that there are female pharmacists who have traveled this low road path, but I can't speak for them. I am not a woman and, unlike a few men, only one woman has confided in me. She actually threw herself to the shrews. She left her husband and two children for the band director at the local high school. She actually did an about face about her job as a pharmacist. It became her refuge. She ended up paying both child support and alimony. What other pharmacists perceived as horrors were tolerated very easily by this woman. What she feared most was loneliness and not being able to make a living so she could still be a good mother every month when she wrote out the checks. She expressed that she was surprised how giving money to her ex-husband soothed her guilt. She did not think that was possible. She finally understood how so many men felt.

I really liked this woman and spent some time talking with her when she needed to talk. We did it on the telephone because my second wife expressed jealousy. This woman told me everything. The affair began when she missed most of a band concert because she had to work. Her daughter iced her out and refused to talk with her about it. She confided in the band director. He invited her to have a drink with him and, since her husband was out of town, she did something she would never do if she was not so emotionally beaten down. She accepted.

She was a Prisoner at work and a Prisoner at home. She put up with abuse at work and abuse at home. Her husband loved the money she made but complained about her job. He was a bully and the fact that she made more money than he did just exacerbated his meanness. She couldn't do anything right. If she prepared meals in advance, they were never what the family wanted to eat. If she did not cook, she was accused of inattention and neglect. If she did not clean the house, the husband complained. When she hired a cleaning woman, the husband said that she was wasteful. She couldn't win at home and she couldn't win at work. The pity was that she believed that this was a normal marriage relationship. One sided. The pressure was too much and what eventually broke down was not her job. She kept working, but what fell apart were her wedding vows.

The band director was not especially handsome or charming, but he was attentive and supportive. He just listened, she said. At the third furtive meeting at the cocktail lounge of a hotel in the next town, the new couple graduated to a guest room upstairs. She had never been with any man other than her husband up until that night. She told me that she was at first uncomfortable because the band director was so gentle and so patient. She was used to a quick, grunting business and never expected to get much pleasure from it. That changed that night.

I will never excuse cheating, but I can never justify a husband who is a boorish brute either. So, I'm not going to be a judge in this matter. I will just say that tolerating a brute, be he a husband or a store manager is never a requirement of a job or a marriage. Never mind about the husband. I can't vote about this woman's husband because I have never been a wife. I love this woman like I would love a sister, but I don't doubt that she embellished her stories about her husband, but I don't doubt for one second the substantive facts were true.

I have been a pharmacist for a long time, however, and I reserve my right to vote about a store manager who mistreats a female pharmacist. A man with a little bit of power can be like a hyena smelling blood when he finds that a female pharmacist who makes more money than he does has some weaknesses. Some managers will act like a Jailer with any pharmacist. Can you imagine with what kind of zealotry he can go after a woman who shows any kind of weak fragility?

Imagine how a real male Jailer would act when his real prisoners are women. Now imagine that the Jailer is a drug store manager and not that satisfied with his own situation and needs a scapegoat. Imagine that his own wife is not that happy with him. Imagine that his children give him back talk and that his wife defends the kids. Imagine that his mortgage is too hefty and that his wife quit her job as a clerk at an insurance agency and she still expects him to pay for the children's extracurricular activities. Imagine that she turns her back on him and calls him a loser when he rolls too close and tries to kiss her on the neck. Imagine what he sees when he looks at the female pharmacist who has never stood up for herself. He sees red meat to be pounced on.

Imagine that all of what I have described is true. Imagine that the poor bastard has been emasculated. That's too damn bad. I say pull the plug and let him go sell cars. A drug store manager who mistreats any pharmacist gets my vote to be put on the no call list when the bonus checks are handed out. However, I love women and I was trained by my mother to do my best to protect women in need.

For any manager who mistreats his women pharmacists, I have one vote and that is off the island for good. A manager who mistreats any vulnerable woman is the quintessential pig.

That does not diminish for one second what pharmacists have to put up with in a Prison situation. We work strange hours, often until late at night. We come home enervated, hungry for both food and some company. Often, we find neither and have to fend for ourselves. We live in a cultural hibernation. For some of us, the only real social life we have is through our children. We go to their games when we are off. The school plays and concerts are important so we make an effort to arrange to be off. Other than that, we don't do much.

The best advice for any woman pharmacist, or any male for that matter, is to always look your best. Appearances are everything. If you look worn out and too tired to do your job, it will be easier for the store manager to mistreat you. Wear fresh clothes and make sure that your white jacket is laundered frequently. Don't give the Jailer anything to use against you. He will have no trouble finding ammunition. You don't have to help him. Get your hair cut when it is too long. Follow company policy on facial hair and piercings. Wear adult shoes. Sneakers are okay, but the white ones with springs on the heels look like children's shoes. Walking shoes are good, all one color.

I was hospitalized with polio in 1951. Because of that illness, for decades I have limped when I get tired. In the mid 1970s, before sneakers were socially acceptable for the pharmacy environment, I wore hard soled oxfords. My limp at the end of most days was noticeable. The night before I went and bought a pair of black sneakers, the store assistant manager stopped me at the door. This man was destined to become one of the most rigid store managers I ever worked with. He stopped me and said, "Jim, you are really limping badly tonight. Are you sure you can do this job?"

I was taken aback and actually stuttered trying to say, "Of course I can do the job." Immediately, I became concerned. Could they take the job away from me just because I limp when I am tired? Would my district manager stick by me? Would the store manager make my life miserable when the assistant told him about this? I worried unnecessarily about this for days. I was institutionalized. I was a Prisoner in a Prison environment and the two managers were the Jailers. It was totally unreasonable for this incident to ruin my peace of mind, but I could not see that. A prisoner in a real prison, with real jailers would tell you that they know exactly what I went through.

In the early 1980s my family belonged to a California swimming pool association. We joined because my step-daughter wanted to be on a swim team. We used the pool for swimming a few times a week and spent time with books in the sun or in the shade under the trees, but the swim team was the anchor. It is pathetic, but the swim club, swim meets and work were my life. I rarely went to the movies and I love movies. It was even more rare to invite my wife out on a date for dinner. We lived in one of the most dynamic spots on the planet, the San Francisco Bay Area, and we may as well have been living in a provincial backwater fifty miles from Idaho Falls. I was culturally deprived by my own lack of interest in life.

I love the ballet, the symphony and the theatre. The last show I saw was A Chorus Line in 1975, the New York cast in San Francisco and I had to be dragged to that. Before that, it was Hair in 1969. I was a Prisoner before it was fashionable for pharmacists to be victims. I believe that I was institutionalized before there was an institution. Perhaps, I am not alone. Could it be that pharmacists are emotionally suited to this type of life? If they were otherwise, would they have become doctors?

There are those among us who have a devil-may-care attitude about it all. They are the ones I looked down upon for years. They didn't seem to care about anything. They didn't care about all of the markers that made a chain pharmacy successful in the eyes of the Wardens.

Back in the day when one hundred dollars for a bottle of anything was a lot of money, a relief pharmacist bought a new antibiotic for over one hundred and twenty dollars for a bottle of 100 tablets. The prescription was for twelve tablets and the doctor was from out of town. The chances of using this medicine before the expiration date was between slim and none. I asked him why he ordered it.

"Because the customer has a prescription for it."

"But it will go out of date and we'll lose a lot of money. Why didn't you just tell him that you couldn't get it?"

"Because he needs it." He looked at me. "It's not your store, man. They sure have your soul, don't they? You are a pharmacist, man. Don't forget. You are a pharmacist." He watched me for awhile. I must have looked lost because his tone softened. "You take this too seriously. They are not going to fire you for practicing pharmacy and if they did you can find another job tomorrow."

"I like this store. I have been here a long time."

"You like it too much. You are habituated. It's like a drug. Maybe you should just quit and go out and find your soul again."

I don't think any of these pharmacists who were actually free from the institution were pharmacy managers. Some of them had been managers in the past, but had bailed out for one reason. They were not willing to put up with the artificial demands of the business. It certainly was not because of professional reasons, because they were professionals in every sense of the word. It was simple. The job depressed some of them. They stepped down and became staff pharmacists.

One man was single and said that he got so sick of the company run around that he found his dream job after he quit the manager position and took a job with a grocery chain. He was a floater in a rocky mountain state. The towns were far apart so the jobs were far apart. There were days when he would work until nine at night and be expected to show up at nine the next morning at a store two hundred miles away. His shifts were always twelve hours and he was paid time and a half over eight hours. The company paid him for his hours on the road and paid all of his expenses for five days a week. Meals, hotel, laundry and he had negotiated an additional one hundred dollars a day per diem. When he talked about it, his eyes gleamed. He loved his job and I could see why when he explained the reward for all of this work. It was Labor Day weekend and he told me that he would work only until the first week in November and then he would take five months off and head for a place on the beach in Costa Rica.

"How can you afford that?" I asked him. "Five months off?"

He smiled. "I get tired, but I don't even have an apartment. The company pays for my hotel room every single day I work and that is seven days some weeks. I pay for the others. It's the end of August. My year started the first week in April and I've already made over one hundred and twenty thousand dollars. That's in five months. I have no family." He looked at me and saw something. "Yes, I work a lot, but those twenty weeks off every winter on the beach are golden." His eyes widened. "I mean on the beach. The place I rent is on the beach under some palm trees. The closest neighbor is a hundred yards away. The village is a half mile down a dirt lane. I spend every evening in a little bodega with the locals. We drink Mexican beer. I even have a girlfriend down there." He shrugged. "If she gets bored with me or gets married, there are other girls." He told me that he would be at this one store for at least six weeks. The pharmacy manager had fallen off her horse and broke her leg. "Compared to my days as a manager, life is good." He winked. "Very good."

To a person, every pharmacist who was happy with their job as a pharmacist had, at one time, been very unhappy with their job. They seem to be able to do the job their own way and to not give any consideration to the expectations and demands of the institution. Most of the happy pharmacists have gotten to the point where they don't care if they are reprimanded for not making the numbers. They practiced pharmacy at their own rate in a safe manner.

One man in Vermont takes a half hour during a nine hour shift to have an uninterrupted meal break. He leaves the pharmacy and takes his brown bag to the office. He puts in the ear pieces of his iPod and listens to music as he eats a leisurely lunch. He reads the paper and always calls his wife to see how her day is going. No one has ever criticized this behavior after he made it clear that the only way he would quit taking a half hour lunch is if they fired him.

I work with a pharmacist who routinely closes the drive-through when the evening rush comes. So far, no one has complained about it. A woman in Washington State unplugs the telephones because she knows that if she answers the phone she is liable to get waylaid by a patient who just wants to tell a story or ask questions incessantly. She figures that a patient pissed off about the phone not getting answered is better than a livid patient who will complain about her being rude because she has to blow them off.

There is a pharmacist in California who got sick of the restrictions and rules of retail and now makes his living as a psychiatric pharmacist. He has the authority to write prescriptions after the psychiatrist or psychologist does the diagnosing. He put in many hours to become certified to do this job and he loves every minute of it.

These people understood that the conditions that were imposed on them in retail were not normal for a professional. The conditions are, however, very normal for the institution of retail pharmacy. All of the things we hate so much are normal. The ridiculously long shifts that are required at some stores. The lack of rest periods, for meals or just rest. The constant supply of beef jerky, sugary candy or salty snack bags on the counter with a steady diet of caffeinated beverages. The standardized equipment that is an ergonomic nightmare for 80% of the pharmacists. The fight a pharmacist has to engage in just to get a vacation in the summer.

Hours and hours on our feet when we know that our legs will eventually be worn out. Bullying store managers. Sexual harassment. Age discrimination and gender discrimination. A condition where a tired pharmacist is dangerous and where every day the health of patients is in peril. All of this is normal.

A psychologist once said, "There are things which must cause you to lose your reason or you have none to lose."

In many cases, our reason is absent after we become institutionalized. An abnormal reaction to an abnormal situation is normal behavior. All of those things that we hate so much are perfectly normal because we are institutionalized and because we tolerate them.

New pharmacists are often preoccupied with a sort of longing for the ideal of professionalism. Then they become disgusted with the reality of what it really is. You can't continue like that. There has to be a killing or mortification of normal reactions to the conditions. The pharmacist becomes numb or anesthetized in order to survive. In the beginning, after being institutionalized, the pharmacist will look the other way when a fellow pharmacist is treated unfairly. Later, he becomes a disinterested, helpless observer. The pharmacy manager is about to be written up again for being over his budget on technician hours. The staff pharmacist does not care, feels nothing about this. It is just the way it is. I worked with a pharmacist who had limited skills, but he worked very hard. He showed up on time and did his job. He was slow, however. When the store manager fired him, I was called to the office to witness the firing. I watched, without any emotional upset, the scene that the store manager had prepared. It was a hatchet job. The fired pharmacist left the office and the new pharmacist that the store manager had hired walked in. I felt nothing. Life went on. I didn't feel shame about that until years later when I happened to run into the guy who had been fired. He was working for an independent. He asked me why I hadn't stood up for him. I had no answer. On retrospect, my lack of emotion surprises me. I am not that kind of man. I was institutionalized, but is that a good enough excuse?

The necessary protective shell is made up mostly of apathy. I keep returning to this theme because it is central to the pharmacist's predicament. The blunting of emotions and the accompanying feeling of helplessness is what allows for abnormal situations to be accepted as normal.

When I was a manager, I was paid a salary. Our base hours were 40 hours a week. We were short staffed and had difficulty getting the warehouse order on the shelves in a timely manner. Even if my shift did not start until the afternoon, I was in the habit of coming to the store and marking and shelving the order as early as six in the morning. It was a foolish, childish loyalty to the job. Oh, what a good boy was I.

One morning, there was a knock on the pharmacy door and I let the store manager in. He asked me what I thought I was doing. I told him that the job had to be done. He laughed at me. He told me what an idiot I was for putting in extra hours for no pay.

What really bothered me was the injustice of his assessment. It was unfair of him to belittle me. It was the worst of insults. I thought that I was being a good manager and I became the butt of a joke. Other employees knew that I did this, but it had not been a comic source of laughs until the store manager made it a funny story at my expense. I was institutionalized.

It is interesting to me that I don't know very many pharmacists who are religious. I would expect that people who feel in a hopeless situation would look for guidance and comfort in organized religion. There be many, but I don't know them.

After working in these conditions for awhile, reality dims and the overriding efforts becomes centered on one task. Getting through the day. It is typical to hear pharmacists who work the early shift to sigh with relief and say, "Well, another day is over." I don't think that I ever asked any pharmacist, "If it is that bad, why don't you quit?" I knew the answer and I didn't want anyone to ask me why I didn't quit. I was a slave of comfort and was fearful that I would have to give up too much if I quit or was fired for refusing to comply with the rules of the institution.

Fear is not a good reason to do or not do anything, but fear is a powerful motivator. If you fear that you will not be able to provide the basics for your children, you will toe the line. If you fear that you will lose that nice new Lexus, you'll tolerate the day to day indignities that all pharmacists put up with and are part of the job anyway. If you fear that you will have to give up some of what you consider to be the nice things in life, like the season tickets, the membership to the club or the new boat you'll let your ideals bend. If your spouse depends on your salary to achieve a certain status in the community, you may fear losing that salary. You will notice that we are discussing a fear of loss.

I was a relatively well off pharmacist in the late 1960s and early 1970s. California was a brain-drain state and pharmacists made more there than in any other state. My wage at my first job in California was 270% larger than the wage at my job in Ohio. My first wife loved the money and she loved the status that it brought her. She had more money to spend as she wished than any of her friends. The one who was the wife of a Ph.D. Chemist for Dow Chemical was green with jealousy. I was young and I took some pleasure in this. After a few years, however, the marriage was not working out and I realized that if I could not continue making the big money, my wife would be unhappy. It followed in my thinking that if she was unhappy, she would go out and find someone else with money to have an affair with. My job became more about conforming with the program to make sure that it was secure than about practicing pharmacy. I stayed in a job I did not like because I feared what would happen if I didn't bring home the big paychecks. I was living in fear of losing my wife. She did have affairs, by the way. She had affairs with many rich men and it didn't matter how much money I made.

In the end, it is fear alone that keeps us Prisoners of Comfort.

You gain strength, courage and confidence by every experience in which you really stop to look fear in the face. You are able to say to yourself, "I lived through this horror. I can take the next thing that comes along." You must do the thing you think you cannot do.

Eleanor Roosevelt

The situation that thousands upon thousands of pharmacist find themselves in is not hopeless. If they see themselves as victims who are powerless in their own lives, they are most likely going to be miserable. However, if they see that they have choices at every turn along the way, there will always be hope.

I have done my best to illustrate the worst pit that many of you wallow helplessly in so that I can show you that you are not helpless and that the difference is a matter of perception, pride, dignity and integrity. Pharmacy will not make you proud. Your viewpoint of yourself and how you behave is what drives your self-respect or lack of such. It is not easy to intimidate a person with confidence and self-respect. Especially, a professional. The relative intelligence and education of the average Jailer cannot compare with yours. He knows that and does everything in the book to keep you controlled.

The sense of being powerful in our own lives is largely missing with the institutionalized pharmacist. Like a disaffected teenager, we compensate for a lack of power in our professional lives by seeking power in ways that are not necessarily healthy. For every female pharmacist who finds some power by volunteering at the hospital for children with neurological disorders, there is a pharmacist who is depressed about her life and stays at home on her days off, lying on the sofa, watching soap operas. She is so used to salty snacks that she can't stop. There is an empty large bag of chips on the floor and a half filled 2 liter bottle of soda. She has gained weight and hates herself for it.

For every male pharmacist who is on the Board of Directors of the local school board, there is a man who gets his power the way I did, from controlling my body to the point of distraction. I cross trained every single day. I spent an hour or more on the Nordic-Track machine six days a week and swam laps four times a week. I restricted calories to the point that I was so lean that I actually looked like I was ill. A friend of my brother expressed his condolences. He said, "I'm sorry about your brother." Mark said, "Why? What's the matter with my brother?" The friend said, "He's got AIDS, doesn't he?"

What I got was power over my own life. I may have been a victim at my job, but here I was in control. Nothing was more important than my physical condition and my lean appearance. I cringe when I see pictures taken back in the day because I actually believed that I looked good.

Recovery is regaining Your Power

Essentially, the miserable pharmacist is wretched because they choose to be unhappy. There is a choice every single day to be proud of what they do or to blame the job because they are not happy. They don't even use the best tool available to them. That tool is anger!

Anger is fuel. It is not the bad thing that your parents said to suppress as mine did. "Jimmy, nobody needs to know you are angry. You should control yourself." We feel anger and we become frustrated when we hide it because we want to do something. This goes against the image of the calm, in-control professional. Instead of showing the anger, we stuff it and chug Maalox and take two 20mg omeprazole every day.

How would it look if we showed that we were angry? At work, you don't hit that someone or break that something or throw that fit. If you smash that fist against the wall, it is in the bathroom where no one can see that you are out of control.

What we do with our anger is deny it. We stuff it so far down that we forget what makes us angry. We are institutionalized and we know that we should not get angry. We lie about being angry at the store manager. We hide our anger at the lack of technician help. We do not express our outrage to the district manager. Doesn't he know that it is his precious customer service that pays the price?

Some of us hide it so well that we medicate the anger and filch the occasional lorazepam to hide it even better. We are professionals and professionals are nice people. We bury our anger. We block it and we hide it.

What we do best with our anger is lie about it. Unfortunately for our spouses, we lie so well that we often take our misery out on the people we love (or are supposed to love) the most. We do everything but listen to our anger.

Listen to your anger. That is what it is meant for. Anger is not a polite request. Anger is a scream. It is a command. It is a slam of the fists down on the table demanding your attention. Anger has a right to be heard. Anger should be appreciated and valued. Anger must be listened to if you are to regain your professional balance and power. Why? Because anger is an atlas or a chart or a diagram back to living the ideals you had when you were in pharmacy school.

Anger reminds you of your boundaries and limits, the areas where no one was allowed to tread without your permission. If you can set up the periphery of your professionalism in just one area, more will follow. If you list only ten serious drugs that you will counsel on no matter what, your list will be twenty in little time. If you let the store manager know in writing that his touching you at anytime, in any manner, is unwanted, you will regain enormous power and control over your own life on the job. You can gain power simply by refusing to get wet underpants because you neglect going to the bathroom when you have to go. Documenting anything at work that makes you uncomfortable will give you surprising control.

Anger shows us where we want to go. We may not know exactly what we do want on the job, but our anger tells us, without ambiguity, what we sure as hell do not want. That is a really good place to start because anger shows us where we have been and sets us on the course of recovery. Anger is not a sign of disease. It is a sign of health. If you no longer get angry at being institutionalized, stop, take a deep breath, and examine how you will find your way back. I contend that you will find that the first sign of recovering your health, well-being and pride will be anger. Welcome it. Savor it.

It is not very healthy to act out from anger. That is childish and not productive. I quit a job once out of anger. It was a good job. I was well respected in the community. The problem was that the store manager tried to micro-manage my department. I have never bent to management from a non-pharmacist. This guy was out to bring me to my knees. I fell right into the trap. I became so angry that I brought the problem to a head with some stupid brinksmanship. My district manager did not back me as fully as I wanted, so I quit. My one-way commute for that job was less than ten minutes. The one-way commute for the next job was ninety minutes. I was like a teenager having a meltdown. I turned my anger into indignation without any examination of the circumstances. I was an idiot.

Anger is there to be acted upon. Anger points the direction. Anger is the wind for our sails as our sailing ship tacks as we move on the appropriate bearing where our anger guides us. Had I used my head and had the presence to translate what the anger was telling me, I would have made better choices.

"Damn it, I could run a better pharmacy than that!" This anger says that you want to have your own pharmacy, you just need to put all of the pieces together.

"I can't believe it. Mildred told me that she was going to demand a transfer to the suburbs and she got it. That's what I wanted." This anger says: Stop keeping your goals and dreams hidden.

You need to express your wants and believe that you deserve your dreams to come true.

"That was my idea. This is unbelievable. I mentioned it only once and that son of a bitch took my plan and put it to work. He gets all of the credit and I get none." This anger says that it is time to take yourself seriously and show yourself some respect. Your ideas are good enough to do something about.

Anger is the tornado that blows away all of the restrictions and hesitations and lack of self confidence of our old lives. Anger is a valuable instrument to be used productively. Anger cannot be the master, only the servant. Anger is a deep well of power, if used properly.

Apathy, laziness, misery and gloom are the enemies. Anger is not a good buddy, but anger is a friend. Not a mild-mannered friend, but a very loyal and steadfast friend. Anger will always remind us when we have been cheated or cheated upon. It will always tell us when we have been deceived or when we have betrayed ourselves. Anger will tell us that it is time, finally, to act in our own best interests. Anger is not the action itself. It is the action's invitation.

Watch out what you ask for

You might just get what you want and then what are you going to do? It can be scary, having dreams come true. That means that you have to take responsibility for your own life. This is not comfortable, but you will feel the power. You can no longer blame the big bad store manager wolf for your lack of integrity. You can't say that the company made you do it. You are back in your own hands, a professional making choices every day that benefit you and your patients. This is a good thing, don't you think?

When you take responsibility, things happen that you cannot fully understand why. You are the pharmacy manager and you tell the

Jailer that you are going to do what is best for your department in all business and professional matters. You tell him that you are the pharmacist, that you know best and you request that he mind his own business. You take all appropriate actions, the department thrives in all areas. Your pharmacy is suddenly the most professional and most profitable in the company and everyone wants to know why. What did you do?

Taking responsibility is not easy. You can feel very much alone. It takes courage to do the right thing. This is a difficult and slippery slope. A pharmacist who has little self-respect and has been stripped of dignity may need assistance in making the choices that are best for both their professional and personal lives. I honestly do not think that you should rush. You have been institutionalized for years. There is no hurry. You don't want to make a rash move as I did. You probably should not try to do this alone. Talk to someone you trust before you take any big action. You can write to me.

I merely took the energy it takes to pout and wrote some blues

Duke Ellington

There is a way back

Deborah Van Sant is a pharmacist in Arizona. This is a snow bird state with lots of pharmacies competing for the baby boomer trade. Deborah Van Sant opened her own pharmacy in 2010. She told me, "If I only had one word to describe how I'm feeling about my choice to open my own store, it would be LIBERATED."

"I have a license to practice and I am now using that license as I have been trained to do. I no longer have to compromise my ethics. I no longer have to put profits above safety. I no longer have to work 14 hour shifts!"

Deborah gave a talk to more than one hundred students at the University of Arizona, College of Pharmacy. Most of the questions at the end were not about the subject. The students wanted to know all about her opening her own store. Deb's store was suddenly a sought-after rotation site.

Deb said, "The chains are our best advertisement."

Nathan Schlect owns a drug store in North Dakota. He does very well. North Dakota requires that pharmacies have majority owners that are pharmacists. There are no Rite-Aids, no Wal-Mart Pharmacies. No pharmacies in grocery stores. Enough said.

RxJoe owns a pharmacy in Boston that is so close to a Rite-Aid that you can read the sign from Joe's front door. There are two CVS stores within a mile. One where he was the pharmacy manager. He started his store with a compounding laboratory. He had to work hard in the beginning. Long hours. RxJoe is a good businessman as well as a good pharmacist. He told me that his compounding laboratory and unit dose business for nursing homes are the winners.

Rich and Marge McCoy are a man and wife pharmacist team. They own the Lopez Island Pharmacy in the San Juan Islands of Washington State. The store is a full-service drug store in a small village in one of the most beautiful and idyllic locations on the planet.

I happen to believe that opening your own pharmacy is not just an airy-fairy idea. I believe that, at this time, having your own pharmacy is a viable and con be a profitable alternative to working for your living at a grocery store, big box store or one of the chains. Pharmacists, by tradition and by law, are the medical professionals who dispense prescriptions. We are always there. It's the law.

Think about this: In 1999, Americans spent 104 Billion dollars for prescription drugs. Prescription spending increased to 234 Billion dollars in 2008. That was an increase of 130 Billion dollars in just nine years. An increase of 125%. Moore's Law of exponential increases tells us that it will not be that long until pharmacists fill One Trillion dollars worth of prescription in one year. That is one gigantic pie. Why shouldn't you get some of it?

Among people older than 60, the researchers from the National Center for Health Statistics reported that 88% are using at least one prescription drug and 67% are taking five or more prescription drugs. That's today's figures. How many will take ten or twelve in Rx drugs in five years?

The baby boomer represent the largest and richest consumer group in history. The first boomer to get Social Security pension payments filed on October, 15, 2007. The baby boomers will not age gracefully as the generations before them did. They want to remain youthful, strong, active and sexy for the entire ride.

Big Pharma is committed to helping them get what they want. Viagra was the first drug designed for the boomers. It is for relatively healthy people. The bank is choking with money that the boomers will spend on their vitality. Watch for seligiline in low doses every other day. A doctor in The Czech Republic touted seligiline as an anti-aging potion twenty years ago. It is already all over the Internet. Big Pharma sees the opportunity for landmark profits serving the baby boomers. Big Pharma will be driving the train. In 2010, they began to promote testosterone replacement therapy to men by advertising a condition they called "Low-T" on television. Why fight it and tell me a good reason why you shouldn't get a slice of this pie?

It is conceivable that a pharmacy practice based on serving only the boomers will be able to provide a good living for the owner. However, no good businessman would exclude any group. The population of the United States is over 311 million and growing. Pharmacists may see that the need for prescription drugs may be driven artificially by the manufacturers, but are you going to go on a Don Quixote mission? Are you going to tell the mother of a member of the 6% of all American children who use drugs for ADHD that the condition is a myth and all they need to do is give their child the attention they need. Are you going to let someone else fill those prescriptions? Remember, Big Pharma will be driving this train.

I don't think that an independent pharmacy will fill the bank filling prescriptions. Not with the ridiculous contracts the PBMs demand these days. There is a lot more to be done.

Let's focus only on the baby boomers. They will have needs that will be begging to be filled. A niche or "boutique" business that the needs of the aging Americans will thrive. Nutritional support is particularly important to an age group that cannot get all the nutrients needed for good health from their diet.

For starters, the baby boomer women have needs that are not being met by anyone. These people get perfunctory attention from their doctors. With a little bit of education and planning, a female pharmacist could become a Women's Health Counselor. A woman's plumbing is not what needs attention. That is well attended to. What the baby boomer woman wants is help with hot flashes, loss of short term memory, lack of libido, loss of vaginal integrity, lack of interest in life, sleep disturbances and depression. Menopause is difficult. The woman is no longer young and she knows it.

Compounding can be a niche business. I am acquainted with a pharmacist in Vermont who does all sorts of compounding in his laboratory. He specializes in bio-equivalent hormone replacement therapy. He sends saliva samples to a laboratory in Oregon, analyzes the results and consults with the patient's physician to determine the appropriate formula. Giant Eagle Grocery Stores recently purchased an independent and will make it into their compounding pharmacy. Why does that bother me? A grocery store? Walgreens has compounding centers. Why not you?

Making your store a center for immunizations can give you a solid clientele and enhance your professionalism, reputation and standing as a member of the health care community. I stopped into a small, vintage drug store in 1999. It is in the Fairhaven District in Bellingham, Washington. At first look, it was a crowded, old-fashioned store, very old-fashioned. Then, I noticed that he advertised every immunization there is available. He was a front line immunization center ten years ago. What is wrong with that? By definition, pharmacists shall administer pharmaceuticals as well as deliver them.

I am passionate about this, but some of you may not have the interest in owning your own store for any of a number of reasons. Not the least being money. For most of us, opening a store would mean borrowing the money to get started. Small Business Administration loans go for as 3% interest. Think about it. Don't let fear run your life. Allow anger to guide you.

For those of you who want to continue working in the retail world for a chain, grocery store or big box store, my best advice is to insist that the people who manage you be pharmacists. Express yourself appropriately when there is something that is not working or that you do like. Communication is vital to your satisfaction in your job. Calm reasoned communication at every point of contention is much better than allowing something to fester and eat you up inside. Remember that anger is good if it is used to guide you. Don't allow it to run you.

The institution will still be the institution, but you do not have to remain institutionalized. Taking baby steps is okay.

Your Ace of Trump

The laws of pharmacy favor you enjoying a satisfying career. In the end, the laws make you the final arbiter of every single thing regarding the dispensing of prescriptions. Prescriptions cannot be sold if you are not present. Non-pharmacists are not allowed to enter the pharmacy unless you are present and give your permission. There was a time in my career when the pharmacy departments were not locked.

(That still is the case in some stores) The entire store could not open unless the pharmacist was present. No matter how you feel about your job, you are the King or the Queen of the store. Pharmacists are the most important employees that any drug store has. Don't forget that. Without you, they can't even call it a drug store.

You are required to counsel patients on new prescriptions. That is the law. If you neglect this, you are an idiot. Counseling is what will differentiate you from the Advanced Certified Technicians who are coming. If you choose not to comply with the law, you may be categorized as less than a pharmacist who is paid $10,000.00 per month. That's the tough love.

The gentle love is that no person can tell you not to counsel and no institution can be designed to make it impossible for you to counsel. The law is the law. If you are mean-spirited, you might actually hope that some Jailer warns you not to counsel so much, criticizes you for the time you take to counsel or simply tells you not to counsel. If that happens, and if it happens consistently, a written complaint to the state board of pharmacy, with complete documentation, is appropriate.

I have portrayed non-pharmacist store managers as villains. Not all of them are difficult. Some of them are friendly managers who are easy to get along with. Some of them know just how important you are. That being said, they are not pharmacists.

The End

Resources You Might Find Interesting

http://www.jimplagakis.com

You can find a few years worth of Jim Plagakis' column JP at Large at the Drug Topics Website http://www.drugtopics.com

You can also subscribe to have the digital edition of Drug Topics delivered directly to your in-box every month.

http://advanstar.replycentral.com/?PID=301&V=DIGI

JP at Large Up to Date is a compilation of the first 169 of Jim Plagakis' columns that first appeared in Drug Topics Magazine in January, 1989. It is published by Advanstar and is available here:

http://www.industrymatter.com/jpatlarge.aspx

Jim Plagakis, RPh

3100 75th Street #14

Galveston, TX 77551

jpgakis@hotmail.com

www.ingramcontent.com/pod-product-compliance
Lightning Source LLC
Chambersburg PA
CBHW060646210326
41520CB00010B/1759

9 780557 867325